SPOT

NATURE

IN THE OCEAN

by Alissa Thielges

amicus LEARNING

waves

reefs

Look for these words and pictures as you read.

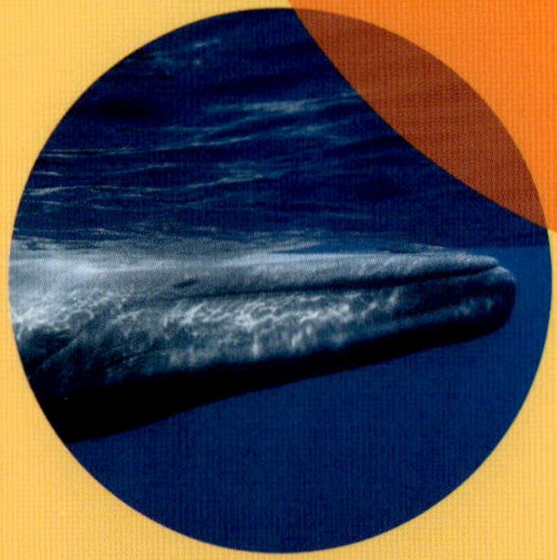

blue whale

shore

What a big body of water!
It is the ocean.

The ocean has salt water.
It covers most of the world.

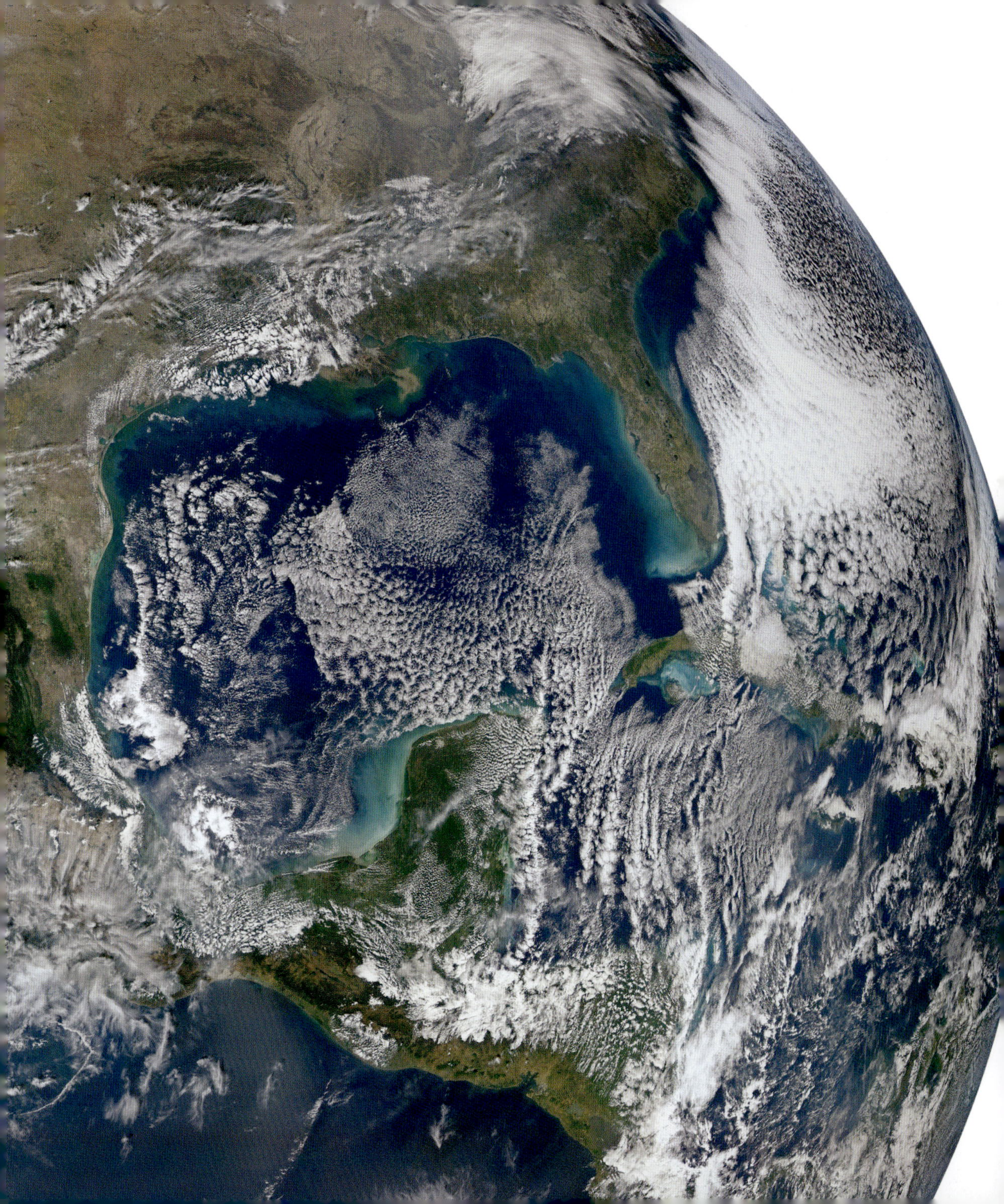

waves

Do you see the waves?
They go up and down.
The wind pushes them.

coral reef

Do you see the coral reef?
It is on the ocean floor.
Fish and crabs live there.

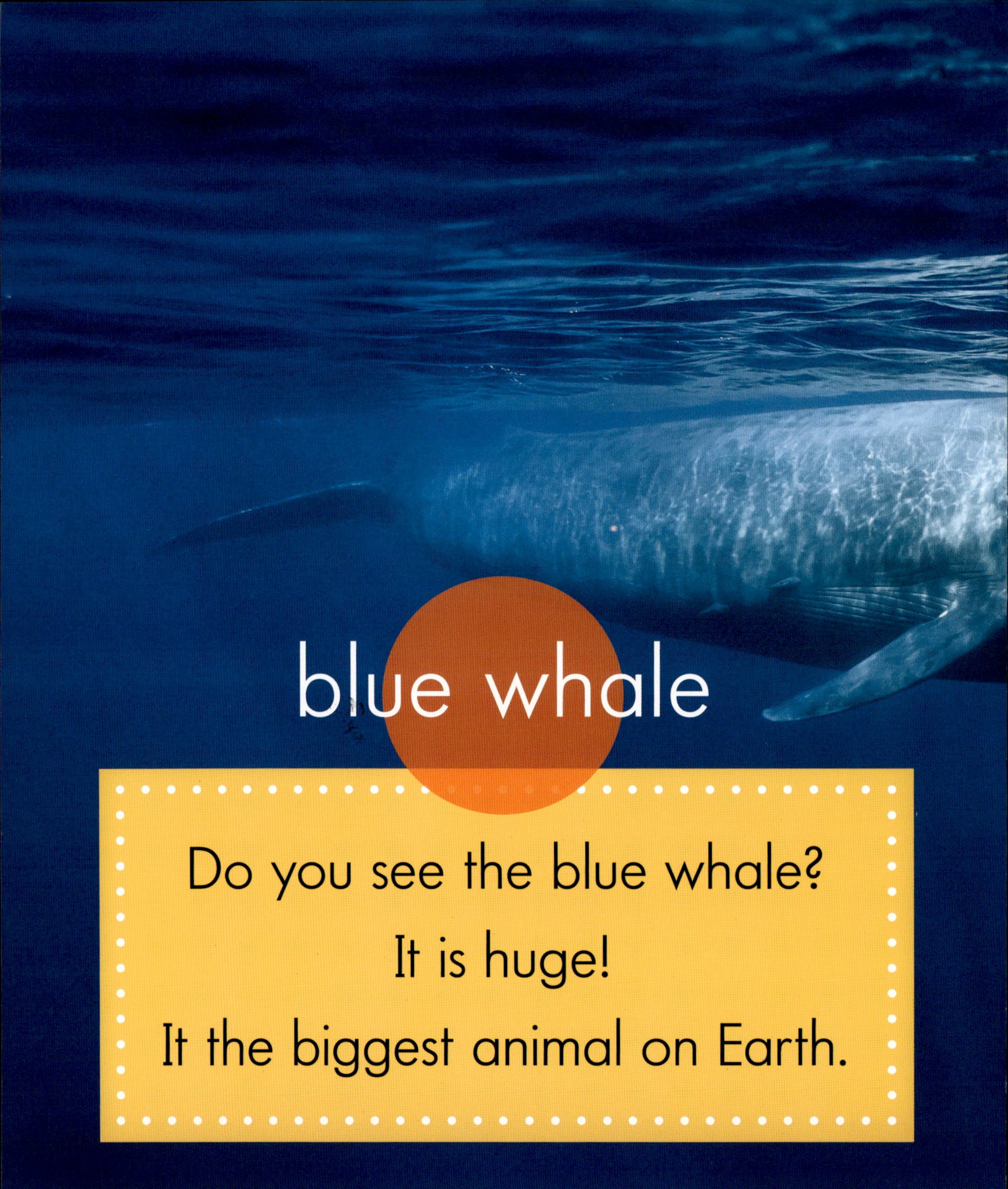

blue whale

Do you see the blue whale?
It is huge!
It the biggest animal on Earth.

shore

Do you see the shore?
Here the water meets land.
It is sandy.

A sea turtle comes to the shore.
She lays eggs.

waves

coral reef

Did you find?

blue whale

shore

Spot is published by Amicus Learning, an imprint of Amicus
P.O. Box 227, Mankato, MN 56002
www.amicuspublishing.us

Library of Congress Cataloging-in-Publication Data
Names: Thielges, Alissa, 1995– author.
Title: In the ocean / by Alissa Thielges.
Description: Mankato, MN : Amicus Learning, an imprint of Amicus, 2024. | Series: Spot nature | Audience: Ages 4–7 | Audience: Grades K–1 | Summary: "A search-and-find book about oceans reinforces new vocabulary to build reading success while close-up images captivate young audiences. A great early STEM book to inspire learning about biomes for kindergartners and first graders"—Provided by publisher.
Identifiers: LCCN 2023011688 (print) | LCCN 2023011689 (ebook) | ISBN 9781645496458 (library binding) | ISBN 9781681529349 (paperback) | ISBN 9781645496717 (pdf)
Subjects: LCSH: Marine ecology—Juvenile literature. | Seashore ecology—Juvenile literature. | Marine animals—Juvenile literature. | Marine plants--Juvenile literature.
Classification: LCC QH541.5.S3 T455 2024 (print) | LCC QH541.5.S3 (ebook)| DDC 577.7—dc23/eng/20230425
LC record available at https://lccn.loc.gov/2023011688
LC ebook record available at https://lccn.loc.gov/2023011689

Printed in China

Rebecca Glaser, editor
Deb Miner, series designer
Mary Herrmann, book designer
Omay Ayres, photo researcher

Photos by Dreamstime/Andamanse, 3, Izanbar, 1, Nataliya Hora, 12–13; iStock/eco2drew, 10–11, ilbusca, 6–7, vlad61, 8–9; NASA/JPPL, 4–5, Shutterstock/Swaroop Pix's, 14–15, Willyam Bradberry, cover, 16